Table of Contents

Why this book?

TLDR

The key takeaways from MOTHER are that there are two main design flaws in the way we organise our societies that are directly or indirectly responsible for the ecological disaster we are in as well as the social incoherence we are experiencing at a large scale. These two are on the one hand an outdated, aristocratic configuration of democracy which is highly vulnerable and on the other a design flaw in the currency system. The easy fix to this cannot be unlocked do to the inherent nature of the problem.

Fortunately, we have all of the technological means to solve the problem anyway. But technology is only half of the picture. The other half is a new way of thinking about economics.

Putting these two together allows us to create scalable, sustainable solutions that have the power to create the foundation for survival and potential thriving of humanity in this century.

OK, Let's begin

You are at school, your next objective is crystal clear. You know what the expected grades are in order to pass onto the next school year. The rules are clear even if sometimes something unpredictable happens, you know exactly what to do to achieve your goal.

You are planning your next holiday. Not only do you know when you will leave and come back, you also know exactly where in the world you will be spending your time, with whom you will be traveling, where you will be sleeping, eating and what you will experience.

You are preparing for your next career step. You know what it looks like and you also know what to do to get there. Usually circumstances force you to adapt your plans, but you know that you will achieve your goal, if you keep your eyes on the prize.

You are living your life. And sometimes, when you have a moment of quiet and reflection time, you even know what you want to do with your life.

You can see it in clear images, you can hear the sounds, you hear the words and feel the feelings. You know what to do with life.

You might not know how these goals got there in the first place, but it is nice to have them - it is good to move towards something tangible, something you can look forward to and celebrate.

Given all of this clarity when it comes to the scope of our personal lives, why is it so hard for us to describe what a good world, a good planet, a good society looks like?

While we meticulously plan our holidays, or next career step, or purchase, we seem to be lacking a picture or an idea of what we want things to look like on a bigger scale.

Why is it that when it comes to grades or holidays it is obvious that the future - in the form of graduating into the next level or a beautiful holi-day - doesn't just happen? Why is it that we don't even question that these things just sort themselves out by themselves, but when it comes to the shape of the future of the planet and society we seem to be happy to let it just sort itself?

People are trying to predict and extrapolate what is happening right now and has been happening in the past and come up with ways the world might unfold, blissfully unaware that this ties our thinking in the past, blissfully ignorant to our knowledge that the future of the world - and anything else - doesn't just happen. Instead, it is something we create, wilfully and consciously.

Of course there will always be circumstances and things that move into our way and seem to be obstacles, but without a clear view of what we want to achieve, we will not even see these as obstacles or sometimes even opportunities, as they arise.

Which is why we need to develop a clear idea and vision and image of what a world looks like with ten billion people living on planet earth peacefully and in balance with nature.

Which is why we need to create a compelling story, a narrative that goes beyond solving problems - one that has the potential to inspire people to create a good planet, a lifeserving society and economy, a place you would want your children and grandchildren to live in.

This little book is an attempt to look beyond the extrapolation of what we know about the world. Instead, it tries to imagine what a world looks like in which all of the problems we are facing on an ecological, societal, human, planetary, economic and ethical level are solved.

A world that is worth building. A good world.

Interestingly such an idea elicits feelings of adversity in many people. People call it out as an unrealistic, dreamy, hippie utopia while taking the much easier to imagine opposite version -a dystopia - and call it the realistic approach.

From the perspective of evolution this is understandable. It is important to be able to imagine all the ways something can go wrong in order to be prepared when things actually do go wrong. This is very helpful in any situation that can be directly influenced by the people involved, by running away, shooting at the sabre-toothed tiger, climbing on a tree or some other lifesaving action. In other words: those humans who could imagine a sabre-toothed tiger to eat them are our ancestors, those who thought he wanted to cuddle were removed from the gene-pool.

Applying this type of thinking to the world and the future of the planet and society at large results in the very understandable, but also extremely short sighted reaction of securing your own personal situation or - at most - that of the society you grew up in. While this approach might be beneficial for a short time, it will brutally fail when taking the entire globe into account.

According to the research of Don Hoffman, evolution does not favour the perception of full reality, but the reception of "fittness", implying that our brains choose which part of reality is useful to survival and only shows us this part of reality. While this might be a smart strategy on an individual level, it becomes are recipe for disaster on a global level.

It has become impossible for one person, one company, one nation to isolate themselves from the global network, not only for direct survival but also for longterm survival. No-one can protect themselves from a dying planetary ecosystem. We all share the same planet - Mother Earth.

Our brains have evolved to prefer using ways to solve problems that are stored in our subconscious to help us navigate the world. It does that because it is the fastest and most energy efficient way to deal with the problem of decision making. Unfortunately our subconscious does not qualify its decisions. It just takes the input, compares it with the stored ideas and creates an output in the form of a feeling leading to a decision.

Adding new ideas to the decision making process of our subconscious works by repetition. You know this from learning anything between walking and driving a car. At one point you don't need to think about it anymore.

In other words, using and repeating a dystopian narrative as the model for the future of the planet will inevitably create just this dystopia because the image will guide our decision making process. This is why we need a narrative of a world we actually want to create.

We have to convince our minds against their societal and evolutionary programming of creating an image of a harmonic, balanced world. We have to learn how to think in the reality of non-linear and exponential terms in order to draw the image and make it happen.

This is a massive undertaking for humanity and for each human individually. So let us not waste more time and start learning, imagining and building the future we want for our children.

I fully acknowledge this and with the magnitude of the thought experiment I am embarking on, I also acknowledge my own shortcomings and narrow mind. This book is not an answer. It is more like a question. A beginning of a conversation.

What does a world worth living in look like?

When setting out to think about and envision a good world, it is necessary to not only understand the fundamental workings of society and the world, but also to create a way to measure success in achieving and building this world.

In this chapter we will explore a way to measure this while In the next chapter we will look at some of the fundamentals that make up societies.

A lot of the conversation around the state of our societies and what to improve revolves around the question of poverty. This definition is rooted in around the most basic, fundamental human needs like food, water and shelter while it largely ignores other aspects of human nature.

Building on what Abraham Maslow described in his 1942 paper as a hierarchy of needs we can establish a new way of looking at poverty through this lens. In his framework he describes five layers of human needs and motivations, starting with the physiological needs layer which revolves around food, water and shelter. Building on this most fundamental layer, the needs for safety aim at protecting the status quo of the physiological needs and to secure them for ourselves as well as our loved ones. This leads to the third layer of the pyramid - the need for love and belonging. Once our safety needs are met, we strive for fulfilling relationships with people we love and who love us in return. But it does not stop there - we also want to be held in esteem by our communities as well as by ourselves, which is expressed in the fourth layer describing the need for esteem. Once we can contribute to our community in a way that is both meaningful to the community and to ourselves, we develop the need for self-actualisation, the need for the expression of the wealth and beauty of our inner world which again both serves ourselves but also lets us serve the greater good.

While Maslow described these levels as building on top of each other in his first paper, he later added that these layers cannot be seen as discrete and independent from one another, but are interdependent and are correlated with all other layers.

It is also true that much criticism of this basic framework has been brought forward in the decades since its inception, but most of the differences between the alternatives are mostly academic in nature. For the purpose of building in the direction of a quantifiable measurement for creating a good world, either of these frameworks can be used. I decided to use Maslow because his framework is the most well known.

Let us take a brief look at the levels from the perspective of this book:

While it is true that the most crucial and life-threatening poverty happens on the physiological needs layer, we can observe poverty on the other levels in modern society. One very current and prolific example is the the mental health crisis we are facing in western industrial nations. While there are many factors involved in creating a tipping point for such a crisis, it can be interpreted as poverty on the levels of safety, love and belonging, esteem and also self-actualisation.

On the safety layer, we cannot seem to make sense of what is happening with and in our world, of how our supposed leaders don't seem to be able to solve the massive threats we are facing. How we are watching the climate collapse and nations fail with our hands bound and no power to change anything. Watching these, does not increase our feeling of safety - taking our fate in our own hands does.

At the same time our need for love and belonging is being compromised by a flood of self-marketing shows on our favourite social networking platforms. Again and again science shows that watching the fabricated images of the perfect lives of other people makes us feel lonely and depressed, even when and if we know that these images are fabricated and that they are a cry for help by lonely and depressed people. Liking their social media contributions does not sooth theirs or our pain. Spending time with our friends, sharing and listening, does.

It becomes increasingly difficult to play a meaningful role in a society and world that is breaking apart on all levels. We are educating our children and young adults with methods and tools from the 19th century and feeding them information that is outdated and does not serve them. We know this and they definitely know it as well. But performing an act on a stage that is crumbling does not fulfill the need for esteem. Building a solid stage does.

When faced with a world that is falling apart, self-actualisation is furthest from our mind unless we use it as a distraction. But even in this case, we know it is so and feel the emptiness.

As a framework of thinking about the world, we can now look at the human condition of poverty or wealth on the five levels of Maslow's hierarchy of needs and we can decide to create a world in which all humans are equally wealthy on all of these layers.

Of course, this will mean different things to different people, of course this is subject to cultural differences and of course there will be relative differences across regions of the world. But it allows us to not only draw an image of a world we want to live in, it also gives us a system of measuring success in creating this image.

It also allows us to let go of our current models of looking at the world and how we run our societies because it gives us a new perspective into the way we operate. This perspective makes it very obvious to us that we are spending a large portion of our energy, human and otherwise, on a small portion of the human condition.

To put this into a vivid image, I like to describe this using Pareto's principle by saying that we use 80% of our capacity to fulfil 20% of our needs. In other words, most of our efforts in the world are focussed on trying to solve the physiological needs.

Let us aim for a world in which we use equal amounts of our energy and capacity on these five layers in order to make every human on the planet equally wealthy on all these layers.

First principles

The world we live in is our construction. Anything other than the rules of physics, biology, chemistry as yet discovered and still to be discovered has been made by humans. - The way we run our societies, our economies, our families, our religions, all of it.

In this experiment we take the liberty of looking at global society from a fresh perspective, free of the constraints and limitations of the status quo. From this image we can then cast back into today and all its complexities with strong information about what action to take.

In order to set out to describe what a good world looks like, we need to look at some of the most important aspects of how we construct our world. I will discuss why these are important, needed and relevant, go on to describe how our current implementation of solving and organising them looks like and what its effects are. Then, to finally propose a way it could be organised to serve as a building block for a world we want.

I will begin by exploring how we **organise the supply** of goods and services because without being able to supply people with things, we cannot have a coherent society. Next I will explain how we obtain and **manage knowledge**, because without creating and passing knowledge on, we cannot evolve or even maintain the status quo.

I will discuss how we **structure societies** in order to be as useful as possible in the paradigms they live, because without a contract between the humans involved in a group effort like a society, interaction would be difficult to impossible.

Then, coming from the macro-level we will look at the more personal aspects of our world, beginning with how we **relate to people**, because nothing ever gets done without other people and because we are inherently weak alone, we have a strong drive to bond with other people. But we can only participate in all of the above with a healthy body and healthy mind, which is why it is important to explore how we **mind the spirit** and **nurture the body**.

All of these dimensions will be discussed on a very fundamental - some might say shallow level. If this is your feeling while digesting this, you are right. As discussed earlier, we want to apply Pareto's principle and look at the most important 20% of how we can think about society. We will even drive this further and reduce this once more to the question what the most important 20% of the first 20% are.

Or, as Frederic Vester expressed it in "_The Art of Interconnected Thinking_", any complex system can be described well enough with approximately 40 variables.

It might not be possible to describe all edge cases and special modalities in an approach like this. However, it is the only approach that allows us to start thinking about such a massive and complex problem from building blocks that can be broken down and used as first principles for building a good world.

Once we have explored these building blocks, I will attempt to describe a coherent picture of how such a world will look. Then, to conclude the book, propose concrete actions that can be taken today to start building this picture.

Organising Supply

Why we do it

If we put aside aspects like love and personal fulfilment, on the most fundamental level, most of human activity revolves around creating supply to satisfy demand. We need food, shelter, clothes, safety, love and so on and the world is busy creating these things and trying to bring them to us. How does this magic work?

There are three basic components that make up all of the activity around satisfying demand. First, someone needs to get the source material. This can be anything between a rare metal, a banana or an idea. This stuff, whatever it is, very often needs to be, or can be, refined into a higher level product. A banana can become a milkshake, an idea can become a religion and a rare metal can become the device you are reading this on (unless you are reading this on a piece of tree - but the idea is the same).

Once refined, the result has to be brought to the recipient, has to be distributed, so someone might open a cafe to distribute milkshakes, someone else opens a series of places for worship and a third one builds a company that sells devices.

In order to somehow steer all of this laborious activity into something that is useful to society, it is helpful to know how the goods and services flow along the chain between sourcing, refining and distributing.

A great way of doing so is by using money. We can let the result of negotiations between the people involved in a transactions become public knowledge and communicate these prices. We call this "the market" - and the "the market" sets the prices.

Also, it is a great way to motivate people into doing things because as soon as their actions create a stream of money towards them, they can then use that money for something they like. That's why we have Ferraris and other things that serve as a token of public self-reward.

We then start creating a set of theories around the way we organise the flow of stuff and services that are being produced in an area like a nation state or a similar geographic region. We call this set of theories an "economic system" - so far, humanity has created two systems of this kind that both tried to become the globally dominating system: communism and capitalism. As we know, communism went away last century, so we currently only are using one such system.

Without regulation any such system invites people to do things that are not generally useful but create a flow of money towards them. In order to stop such behaviour we came up with a set of rules and regulations about what is acceptable and what is not acceptable. This is our governance system. At the moment we are mainly using democracy, republic, monarchy, communism and dictator-ship, each of which comes in multiple flavours.

While the evolution of an economic system is largely driven by the flow of money, the evolution of a governance system is largely driven by the underlying cultural beliefs of a geographic region.

So, we can sum up by saying the driving forces of the way we construct our worlds are money and our core beliefs about the world and their expression in our daily lives are the economic- and the governance systems.

While all of this might sound and feel abstract, it will start making sense in the next chapter, when we start looking at the effects of this setup.

How we do it (and what goes wrong)

As outlined, one of the two core elements of how we construct our societies is money. Usually, when things go wrong, we tend to blame the two systems that we wrap around our currency system - the economic and the governance systems.

We create a revolution, describe a new and - at least subjectively better - system, topple the symbols and leaders of the old system and enter a new era ... until something goes wrong again and history repeats. Without any doubt this evolution has led to benefits for the people, but at the same time it has not led to any fundamental change of action.

The reason for this inertia can be found in the way we have designed our money. For this argument I will not go into the history of how, when, where or by whom the monetary system has been established. It was an evolution over centuries, and going into these details is a distraction for what I want to achieve, with the perspective I would like to offer.

When we create money, we create It out of thin air. Banks create it in the form of a credit. The amount of money they can make this way can be controlled by fiscal law as well as central banks. These credits usually come with an interest rate.

This process of creating debt with interest rates results in an ever growing amount of money in circulation. In fact, since the 1970's, the money supply has shown exponential growth - up until

2008. In addition to the backing system of our currencies, one of the ideas of mainstream economics for regulating money supply is that organisations simply go bankrupt and are destroying money in the process. While this is the case, this cannot counter the exponential growth of money supply inherent in the system.

So in order to keep the trust in the money, we humans want to know that there is some value that backs the money we are using to run our lives. For a long time this used to be gold, but since 1971 humanity has decided to use the GDP as the backing system for our currencies.

This seems like a good thing - after all, our prosperity and wealth seems to be linked with the growth of the profit of the country we live in, so it must be a good thing if that grows. What we miss when taking this perspective, is that the GDP only measures the profit a given geographic area. For example how much a country can make in a year. In doing so it only quantifies the profit but it does not gather or give any information about the quality of the source of the profit.

This means that anything not explicitly prohibited by law and increases profit, is rewarded and welcomed by society, because it increases GDP which in turn is necessary to feed our monetary system.

And this is exactly what has been happening. We have been externalising cost on the back of nature, the planet, humans and society. We have dumped waste, created products that need to be replaced quickly and can be produced and distributed cheaply, at the cost of the environment.

This system is also an invitation to exploit people, by pressuring them into working more hours for the same wage.

In other words, the things deteriorating on our planet and in our societies are directly linked to our construction of the monetary system and not our economic or governance system. It is true that both of these have the potential to decelerate or accelerate the exploitation, but they cannot stop it.

And since we have linked our ideas of how to organise the world and society to the idea of money and overlook the way money actually works against us, we seem to be locked into a situation where we are literally killing the planet that feeds us.

What it must look like

One way these dynamics could be turned around very easily is by swapping the backing value of our currencies for something we want more of. As described earlier, the money supply continually grows - and so does whatever we back our money with. So, backing our money with something we want more of, will inevitably push us to creating more of it.

A possible value for doing so is the "Happy Planet Index" developed by the think-tank NEF from the UK. This index measures life expectancy, wellbeing, ecological footprint and inequality for each country.

But imagine the leaders of the US, China, Germany, Russia and Saudi Arabia ranking on spots 108, 72, 49 and 116 - there is no data available on

Saudi Arabia - of 140 sitting down to discuss to change their backing system from GDP to HPI. This will not happen, after all the US ranks on spot 1 on GDP globally, followed by China while Germany, Saudia Arabia and Russia are sitting on spots 4, 18 and 12 respectively.

What if there was another way? In 2008 a person or group of people - nobody knows- gave the world a protocol of how to exchange value without an intermediary and without the need of trust between the parties because the trust is part of the protocol and therefor built into the system. I am talking about blockchain.

Unfortunately blockchain has been misunderstood and misused by many people, so before diving into how we can use it as a basis for a solution to organising supply, I need to set some of the records straight.

One of the most immediate concerns people tend to have when talking about bitcoin/blockchain is that it consumes a lot of energy. While this is entirely true, it is not inherently necessary for blockchain to work this way only. The reason for the high energy consumption, again, lies in the backing system. In the case of bitcoin, a lot of calculations go into creating a bitcoin and many people - or their computers - are competing to win a race of calculations. In other words, a lot of money and effort goes into creating a bitcoin which inherently gives it some sort of value and backing.

This is also true for other blockchain systems, but it does not have to be that way. The underlying algorithm does not rely on this mechanism. In fact, in can be replaced by a whole range of other mechanisms and the global blockchain community is working on identifying and implementing the best one, without the necessity for excessive computation. And hence, a non-significant ecological footprint.

A second misunderstanding of what blockchain is lies in the understanding that it is just another currency. While it is true that it has a built in mechanism to prevent double spending - one of the most important elements in a currency system - serving as a currency is only the most publicly visible portion of blockchain.

The true value of blockchain, or "distributed ledger technology", how I prefer to refer to it, lies in the potential to create entirely new and independent economic systems. The process of designing such an economic system is called token engineering. This discipline designs self-contained economies, including incentive mechanisms, rules of transaction and exchange, integration with the goods and services trades, and more.

What this means is that any group of people can now set out to create an economic space of exchange and trade, entirely outside of our existing economic systems. These spaces can address any aspect of life, be it food, energy, transport, real estate, education, literally anything, and completely rethink and redesign the value chains involved to reflect the values of the people involved.

In addition to this it also becomes possible for machines to actively take part in these economic spaces because the contracts that define the space are machine readable, the transactions can be exchanged between machines which serves as the basic operating system for fully automating our supply systems.

We can think of our economies as a system of self-contained cells of economic exchange, largely run by machines, both physical and virtual, which are owned by the people, effectively turning economy into a commons and thereby serving every human equally.

These local economies can be technically interconnected in order to ensure access to any type of product or service to any human on the planet.

Of course, simply connecting supply chains on a digital backbone is only half the picture. After all, much - if not most of the things we need to organise and distribute, are physical things.

With modern technology it becomes easier and more sustainable to produce physical goods. It also becomes easer to transport and distribute them.

On the one hand, synthetic biologists around the globe are exploring ways to create new materials that can be grown in the lab while others are finding ways to print raw material into usable objects. In both cases, the cost of setting up a production facility is largely reduced, meaning that small production sites can be set up across the globe, close to the point of consumption.

It also means that these sites are often multi purpose, producing more than one product. This applies to any field imaginable. People have already numerous things between houses, cars, organs, food and much more, using technology like this.

And last but not least, these ways of production reduce the cost on the environment by using less material, sourcing locally and sustainably in many cases, while eliminating the necessity for transport.

Where transport is still required, the first and last mile of the journey is usually facilitated by some form of electric vehicle like a skateboard, bike or scooter, while the mid-range connections are managed by autonomous passenger drones. These connect to fast moving train-like structures which can travel thousands of miles per hour. And then these connect to a network of rocket bases that allow jumps from location to location around the globe.

All of these modes of transport run on sustainably sourced electricity, except the rockets - since it is physically impossible to create the necessary thrust with electric propulsion. But using sunlight, water and CO_2 from the atmosphere it is easy to produce CH_4 as a fuel for these rockets, making them essentially carbon neutral.

All of these modes are organised on a global distributed network, owned by humanity. The same is true for the land used for production of food and goods as well as space for living. It is owned and managed by humanity through a shared distributed network.

These ideas are still very close to the current model of thinking, in terms of moving goods across the planet in order to meet a certain demand. With the help of technology and ancient knowledge, we can live in ways that do not require these types of transport or communication modes.

Building on the example of the earth-ship movement, it is entirely possible to build low cost, very comfortable, entirely off-grid, carbon neutral, or negative housing. These houses contain sections for growing food which can be extended in off-grid vertical farms to grow almost anything, very close to where it is consumed.

This transition will reduce the need for global transport dramatically, almost eliminating the enormous energy consumption for real estate. The disastrous ecological footprint of conventional farming methods can be eliminated, as well.

Structuring Society

Why we do it

Nothing ever gets achieved alone. We all rely on other humans to teach us, to feed us, to help us. We also rely on the humans around us to play by the same rules we do, by rules we all agreed on.

In 1980 Robert Axelrod held a tournament using various ways of interaction between two parties handling the Prisoners Dilemma, an example in game theory showing that two completely rational individuals might not cooperate, against their best interest.

Each proposed solution competed with every other solution in 200 iterations. In each tournament one strategy came out as the clear winner, a solution authored by Anatol Rapoport called "Tit For Tat" which states that the first move is always cooperative and the following moves echo what the other party does. In other words, if the opponent is cooperative, the answer is cooperative, if the opponent is not cooperative, the answer is not cooperative either.

It also turned out that the "nice" strategies, the ones that had a bias towards cooperation, were more successful than the competitive strategies.

In larger contexts than with only two actors and a single set of decisions, we need to come up with rules that everyone agrees on in order to be able to enjoy the benefits of cooperation on a larger scale, effectively leading to situations in which all participants win.

The general agreement that we are better off cooperating and that we need a set of rules to govern aspects of life and interaction in a way that allows people to cooperate with confidence and freedom, is what I mean when I refer to society in this book.

The way we go about structuring society has changed over time and moved towards collaborative and open models across the globe. The number of democracies on the planet surpassed the number of autocracies around the year 2000 and has been on a steady incline since the 1900's. As a result, in 2019 more than half of the worlds population lives in some form of democracy.

Autocratic systems are unfit for a globalized world because they tend to detach themselves from other systems. This type of detachment has become relatively impossible from an economic perspective and absolutely impossible from an environmental perspective.

For these reasons I will focus on democracy in the context of this book. However, I will challenge the way we configure our democracies at the moment and highlight some better options for the 21st century.

How we do it (and what goes wrong)

I believe that democracy is the strongest form of organising and structuring society, yet the current configuration of democracy we are using is deeply flawed and creates extremely undesirable out-comes.

In the next couple of paragraphs I will describe what is going on with our democracies at the mo-ment and - as with other topics - I will look at it from a perspective that allows me not to lose track of the overall goal and get lost in the details.

The most common model of a democracy at the moment looks roughly like this: every couple of years every citizen above a certain age has the opportunity to vote for one or more representatives of political parties. The winners of these elections usually team up with other strong parties and then go on to make the decisions that need to be made for the next couple of years until the next election.

The first problem with this model is that it creates a situation in which political parties market certain ideas in order to maximise their votes. This mechanism distracts from a lot of the real problems and usually does not allow for root cause analysis and diligent truth-seeking. It creates proxy-disputes for the public to see the varying positions of the parties and people involved. These disputes often have as much to do with leading a country, as running a farm has to do with baking bread.

Another problem is that the people involved are usually overwhelmed by the complexity of the issues they have to make decisions on. This opens a great opportunity for any type of lobby work, or even more worrying types of influence on government officials and politicians. More often than not, lobbyists funded by wealthy individuals or organisations write the words which will later find their way into the legislation, which - of course - will be beneficial to the people and organisations they work for.

Looking back at the situation around money, as discussed in the previous chapter, this means that the political process is entirely dominated and run by money. Firstly, because the party and people with the best marketing - which can be bought - win an election, and secondly, because money can buy its way into the decision making process through lobbyism and bribery.

While these effects are more or less pronounced in the various ways we have organised our democracies, the underlying principles remain the same.

What it must look like

On his final trip as president of the United States Barack Obama went to visit Athens honouring the roots of modern democracy. Maybe he did this to hint at the surprising configuration of the Athenian democracy from todays perspective.

It was void of parties as well as elections. Yes, you read that right - no elections in a democracy. Instead, every citizen had the obligation to be part of one of several governmental bodies when drawn by lot to take part. The political participation in Athens was entirely based on randomness.

The natural reaction of most people when learning this is to highlight their lack of trust in the knowledge of common people and the fear that this would lead to bad decisions. While it is true and obvious that no human will be an expert in all possible fields of life, this is also true for our current politicians. So, randomly chosen representatives would use the same type of support as todays politicians do. They would get advice from experts, communicate with the public and stakeholders in their decisions in order to learn as much as possible about the field of their responsibility and make informed decisions.

In fact, several countries are experimenting with this at the moment. One of the most cited example is Ireland where one hundred randomly chosen citizens gather in multiple sessions during the course of one year to come up with a recommendation to the government concerning various topics. In a similar move Mongolia recruited eight hundred randomly recruited citizens to deliberate about a new constitution for the country as personally observed by Larry Lessig.

While such an approach has not yet been tested on a wider scale than one small geographic region like ancient Athens, or on a vertical like gender politics or similar topics, there is no reason why it cannot work on a wider scale.

In fact, it lends itself beautifully to a system which allows for local engagement to be interlocked with wider and more global engagement and policy making. Again, building on technologies like distributed ledgers as the transport and negotiation layer of such a type of democracy, as along with narrow artificial intelligence as a support system for decision making in highly complex environments, this approach becomes feasible. Naturally, any software used within such a system must be open and transparent, owned by the public - in essence it must be a commons.

What stands against such a change of system is the current interest of corporations to have access to a group of people they can influence easily in order to create the regulatory environment which makes it easier for them to do business.

As we will see in the next chapters this is a very important element since business has direct or indirect interests in almost any aspect of our lives. The underlying mechanics of the economy - the currency system - extends its effects into all of these aspects through the power of companies, especially when they can leverage the political system.

As a result, any way of structuring our society in order to form agreed upon terms of how we interact in the various ways life has to offer, must be independent of the economic structures it operates within.

In the final chapter we will see how this can be achieved, when we start to weave the threads spun in the previous chapters into a more coherent picture.

Managing Knowledge

Why we do it

Knowledge has always been one of the key advantages for survival. There are two basic strategies around knowledge that people tend to follow. One is to simply keep the knowledge for yourself. The other strategy wants to spread knowledge.

While in the short term it can lead to great benefits when keeping knowledge private, history shows that in the long run, sharing knowledge is the more successful strategy. One of the main reasons for this is that those who share knowledge get to shape the way people think about the world, about how they act in the world, how they are.

The science around this is called memetics and has evolved from the first mention of the word by William Hamilton in 1963 and popularised by Richard Dawkins, Daniel Dennett and others.

Building on the ideas of Richard Dawkins described in his seminal book "The selfish gene" we can get an understanding of how knowledge is shared, packaged and distributed. This is possible because memes share important characteristics with genes.

If you take the perspective of a random gene for a moment, it is easy to see that the gene does only care about its own procreation - if it doesn't create copies of itself it will vanish from existence. In other words, the fittest gene is the one that is best at making copies of itself.

Memetics takes the same perspective for pieces of information called memes. Memes are interested in making as many copies of themselves because otherwise they disappear. One great example for a meme is fashion. Every new fashion trend is a meme.

Memes that support the survival of a human or the human species are not necessarily the ones that are best at copying themselves. Therefore it is of critical importance that we as humans understand this and carefully distinguish between knowledge that is successful at copying itself for its own purpose and knowledge that is helpful to the planet and the people.

It also is of critical importance to spread the most useful memes to as many minds as possible in order to provide humans with building blocks to create ever more useful and deep knowledge.

How we do it (and what goes wrong)

The current paradigm of education evolved during the time the British Empire was still dominant and thriving. What was needed in order to organise and maintain such a massive wealth of land and possessions was an army of people literate in reading, writing and basic math.

The idea was to teach as many people as possible to do these things by putting them into a room and exposing them to the material they needed to learn. This would create successful agents for handling the tasks required by the British Empire.

This system was very successful, so when companies developed the need for a specialised workforce during and after the industrialisation, this was the method of choice.

In essence, this is the primary purpose and configuration of todays education system. As we have seen in other fields too, the pressure to increase the GDP is driving businesses to build influence with politicians in order to create the educational system that serves their needs best.

At the same time, the vast majority of parents around the globe are very concerned about giving their children a good education, and rightly so. We are very careful and reluctant of allowing any type of experiment regarding the education of our children. This led us to massively protect the education system, which created a trend to harden the traditional way of teaching along with the traditional content.

In a response to the slow evolution of public education, an industry of private education has sprung up to cater to the needs and requests of the modern world and our modern economy. This industry by definition creates an education gap allowing the wealthy to provide an education for their children which fits the needs of our times much better than the majority.

As a result we are currently creating a tiny elite of people who have the necessary skills to run the world by employing modern techniques and technology. This makes more and more people redundant in the economy, while also accumulating wealth and power within the economy.

Most teachers are motivated by giving the children they serve a basis for creating a good and successful life. However, the system sets them up for failure regardless of their intentions.

What it must look like

We have lived past the point where the job of education is copying a huge chunk of information from the brains of one generation to those of the next. Information and even the pattern recognition in vast amounts of information will be readily available to everyone on the planet in the next couple of years.

What we need from an education system is to allow a new generation of ingenious inventors and curious explorers to find solutions, big or small, solving humongous or minuscule problems at incredible speed and adapted to local constraints.

Why is this important? Because the changes in our environment will be extremely rapid both on the ecological side as well as on the technological side. Dealing with this in a positive and life-serving way requires incredibly agile and awake minds.

Luckily we know how to create the environment for such a miracle to happen. All we need to do is supply children with resources like access to information, tools and materials that allow them to explore and learn about any question or problem they might be facing. Of course, some support by an adult is welcome and helpful, in case they get sidetracked or cannot find an entry point.

A pioneer in this field is Sugata Mitra, who stumbled upon the fact that kids are able to teach themselves almost anything with this mix. All we need to do is to make this the norm. This approach welcomes mistakes as a natural part of learning and values process over output, a core pillar of what is known as the growth mindset, coined and described by Carol Dweck.

While this is great at allowing the individual to create her own toolset and set of expertise, we need a solid way to create a process around scientific consensus between people across the globe. This will be enhanced through cultures and languages based on the scientific method, repeatability and evidence.

While the scientific method has proven invaluable for finding hard information, the process of sharing the findings of scientists is broken in multiple ways. Overcoming this with tools of the 21st century is comparatively easy, by allowing everyone to publish his or her results for public scrutiny on a shared space for knowledge.

In this space every person has the possibility to review, comment and cite any published piece. The strength of a piece of knowledge in this system could, for example, be determined by the number of citations, weighed by the citations of the piece citing it. It would work as a mixture between Googles Page Rank algorithm and the current citation mechanism in the scientific community.

By making this entire process fully transparent and independent of subjective perspective, as is the case in the current modus operandi by publishing through a peer reviewed process, it is possible for heterodox knowledge and ideas to flourish that would normally be left behind or, more accurately, left outside the conversation of the scientific community, based on the decision of a few who want to protect the status quo that has put them into positions of power.

Of course, this again is based on a distributed system in order to prevent it being hijacked or owned by a single organisation or individual. The knowledge as well as the process is part of the global commons.

In order to make it a truly global commons, it must be translated in enough languages to make it truly accessible to the global population. Currently this is still a major barrier. A distributed system would allow for peer to peer translation of all text, assisted of course by modern translation technology in order to make knowledge searchable, a necessary precondition to open access of information.

Having talked about the two new models of organising the existing elements of how we think about education in the old paradigm, there is one very important addition to make.

Building on the ideas of developmental psychology in general and Robert Kegans' framework, it is important to formalise what we currently tag as "life-long-learning".

Essentially this is the idea that the mindset evolves over the course of a lifetime and does not halt its evolution at the end of puberty. We used to assume that the mind has evolved and will basically stay the same for the rest of the life of the average human. But this notion could not be further from the truth as neuro sciences have been able to show, often in tandem with the findings of developmental psychologists like Kegan, but more often following the same.

For those readers who are not familiar with Kegans' ideas, I will briefly outline the concept. Basically, there seem to be three layers of development of the human mind. It begins with the "socialised mind" which is mostly influenced and defined by its immediate social surroundings. This enables a person to be part of a community and to be trusted by members of that community by act-ing in predictable ways.

It is followed by the "self-authoring mind" which begins to look at the memes that are used in the community in order for it to work. Then to make conscious decisions about which ones he or she wants to follow, what to leave out and which new ones to include.

The final part of this trio is the "self transform-ing mind". It is able to step out even further and begin to look at its own mind and behaviour and to make transformational decision about itself.

This concept feels uncomfortable to many people because it introduces an unfamiliar hierar-chy. This is not something measurable like a title, a degree or other levels of hierarchy earned in the current system. Rather it is the same shift in quality that happens when children around the age of four or five become aware of themselves. Maybe you remember learning a new concept as a child, like putting letters into words and then putting words into sentences. There is a distinct moment in which it "clicks" and the concept suddenly makes sense. A similar thing happens when moving from the "self authoring mind" to the "self transforming mind", or any other transition for that matter.

The work required to move from one mindset level to the next is a very personal and individual one, but that does not mean that it cannot be formalised. There is a wealth of knowledge about potential paths to take for that journey, from almost any culture in the world.

Providing spaces like this with different offerings for personal growth throughout life, is key.

Also, trusting individuals with not only the growth work itself, but also with the timing and depth of the work is especially important.

Allowing humans at all stages of life to freely work on self-growth will automatically be beneficial to society since the outcome of such practice always results in an expansion of the mind providing more life-serving activity for those who have grown.

This can be modelled after a movement which happened in the early twentieth century in the nordic countries who took an idea from the networks of Goethe, Schiller and Humboldt which was not implemented in Germany due to the revolution in 1848.

To the people at the time it was very evident that a phase shift was coming in the way society was structured. This shift was moving into a more democratic system which required very grounded and grown up individuals to make good decisions on behalf of society. So, starting in Denmark, and spreading across the nordic countries, a movement began that provided free retreat centres to young adults to allow them to find their personal path.

I will go more into this in the chapter "Minding the Spirit".

Relating to People

Why we do it

Have you ever accomplished anything alone, entirely by yourself? You might think you did, but along the way there were definitely people on the sidelines supporting you, either directly or indirectly. We need other people to achieve anything in life, which is one of the reasons we have evolved to create relationships with other people.

As both Alice Miller and Gabor Mate write and discuss, there are two essential needs for humans - the need for attachment and the need for authenticity.

Both are critical for your survival as an infant. You need to feel and create attachment to the person or people who feeds you, cleans you and keeps you safe. At the same time you need to be able to express your needs without language - so you have to be authentic in order to make sure you get what you need.

Later we learn skills from other people and whenever we practice these, we honour the legacy of our teacher. We never do anything by ourselves.

How we do it (and what goes wrong)

Humans have always exchanged things. It is very helpful for survival to exchange things with other people. This is an important function in a society that still requires humans to perform basic tasks, like driving a taxi or cooking a nutritious meal.

But in tandem with the development of the modern market economy - and maybe even before that - we started getting used to the transactional quality of a relationship. "I bring down the trash and you do the dishes", in some variation we all know this type of deal.

The higher the pressure to survive or to perform in a society, the higher the need to optimise ones time for the survival or performance mode.

However, creating transactional deals to improve your own performance is very tempting.

In fact, being really good at keeping a lot of people around you in a transactional mode is one of the hallmarks of a financially successful person in our twentieth and twenty-first century societies. The relatable argument always is that I can do something with my time that is of more value when paying someone else to clean the house, do the dishes or cook.

The very nature of our currency system is entirely based on creating competition through scarcity. In fact, being able to afford lots of money based, transactional relationships serves as a very clear status symbol.

Once arguments like these start creeping into relationships that should be unconditional in nature - like parent-child or romantic relationships - and turns them into transactional ones, we are losing a large portion of what it means to be human. Because an unconditional relationship is the only way to be truly seen by another person.

Being seen, recognised in our entirety ties back into one of the core survival instincts we have, the need for attachment.

What it must look like

In a world in which most work is done by machines, humans stop having to work in the traditional sense. But they will have both the necessity and the opportunity to do something much more important and meaningful. They will relate to other people. They will show up as full human beings in interaction with other human beings, who also show up in their entirety.

These are bold statements, so let me elaborate on the factors driving this:

One, as mentioned, is the drive to create attachment with other people. This is an instinct that is crucial for our survival. This is why, during evolution, nature has come up with several reward mechanisms when supporting other people, listening, helping. These mechanisms help us thrive personally because they release a cocktail of hormones that don't only feel good, but also make us happy and healthy.

Two, given the additional time we have, we can grow spiritually in the right environment. While large parts of any spiritual path can - and often must - be walked alone, learning from other people is almost inevitable, as well as being challenged by other people in our path. In addition to this, moving up the ladder of mindset leads to letting go of the ego, but more on that in the next chapter.

Three - and this is a crucial point - we will need to deeply understand what we, collectively want from the machines. Sometimes this collective is a local one, more often global. In order to achieve this, we need to collect a lot of information from as many people as possible and then aggregate, consolidate and interpret this information. We can achieve this by interacting with other people, learning about their desires,needs and wishes, sharing ours and discussing where we are in agreement or disagreement and why.

Deeply relating to the fears and desires of another person while being and acting as an independent but loving human, will be one of the most important activities. It will come in different flavours and with different motivations, but it will be the new work.

Nurturing the Body

Why we do it

It doesn't require a lot of explaining why it is important to nurture the body. However, we tend to forget how important a healthy body is while it is healthy. So reminding ourselves of what it means to live in a healthy body is useful when thinking about what the best possible world should look like, because such a world implies that we are healthy too.

In order to work properly on this topic, we need to define what we mean by health. After all it is a very subjective and individual matter. In my view, it is important to distinguish between an impediment and lacking health.

An impediment can be anything that resulted from trauma, like a missing limb or organ to a gene defect.

Health means that relative to the impediments we all carry, we can live in a state of wellbeing, physically, mentally and spiritually. This chapter is about the physical aspect, the body.

This is not about how to avoid trauma or DNA defects, I only discuss how we go about and must go about managing our physical health on an everyday basis. It is more about chronic disease.

How we do it (and what goes wrong)

We tend to only start caring about our health when it begins to fade. This is already the most important aspect of how we think about health at the moment, and what is wrong with our way of thinking about health.

Once we start seeing symptoms of any kind, the body already has been pushed out of balance for a long time. It takes time to accumulate enough disorder within our bodies until the imbalance is too strong to hide.

Our bodies are designed to be healthy, not to be sick. When treated right, being sick should be an exception, not the norm. However, today we see several epidemics in chronic diseases, especially in the so-called developed world, but also in the so-called developing world.

Since our bodies don't get sick by themselves it must be something we do with them that gets them sick. The answer to how this works and what we do is surprisingly easy.

In essence, we are masters at creating chemical imbalances in our bodies. These chemical imbalances have two sources: the things we think, hear and say and the things we eat, drink and breathe. While the connection to food might be obvious, the connection to thought might not be at first, but it is very easy to see it. Many of our feelings have physical expressions. For example, people tend to blush when they are embarrassed, or to cry when they are sad, to laugh when they are happy. Those physical expressions are triggered by chemicals associated with the feeling in question. Extrapolating from this, it is not difficult to intuit the things that happen in your body if it is continuously exposed to a negative environment.

By the way, I use the term chemicals very freely in this context in order to not get distracted by the details of which hormone and which molecule binds to which part of which cell. What I mean by chemicals, are the messengers inside your body that inform cells of what they should do, or not.

Basically there are two ways we can influence the balance of our bodies. One works through the gut and one works through the mind.

Currently our health system mostly ignores both of these elements. Training of doctors does not include ways of using food to heal and to keep a body healthy. Nor are they trained to take the mind into consideration, although this goes against the intuition of many professionals in the health industry. The industry is actually set up to work in an entirely different way.

We can trace this difference back to money. The health industry is required to perform very expensive tests on new medication in order to be able to sell them in the market. They also invest a lot of money in research for new drugs. Both of these expenses need to be earned back by the medication sold. So it is in the interest of the pharmaceutical industry to sell as much medication for as much money as possible.

One way of doing this is protecting formulations by patents. Once these patents expire, the formulation gets slightly modified and patented in its new form. This way it is possible to control the price of any medication for a long time.

A second way of using patents in this way is by buying patents for effective and cheap medication, with the intention of not producing it because it would threaten existing revenue streams and products.

For the same reason it is in the interest of the pharmaceutical industry to keep people in a state of dependence on medication. It is not in the interest of the pharmaceutical industry to heal people and help them live a life free of drugs. This sounds cynical and I don't like having to write this, but this is an inevitable consequence of doing business in the pharmaceutical industry.

So the industry uses its financial power in order to create situations that enforce the use of medication. This can be done again through direct lobbying of doctors, indirectly through politicians, through marketing or creating dependencies - in other words, addictions - in patients.

But the pharma-industry is not the only one playing with our health. The food-industry is contributing its fair share as well. Since it is an industry, its primary goal is to make money. Of course it can only make money if it serves us well. However, this turns out to be the prime benefit the food-industry is exploiting.

Certain forms of energy trigger a reward reaction in our bodies. For example, sugar makes us happy. But it only makes us happy in the short term because most sugar is not helping us to stay healthy in the long term. Our health largely depends on the micro-organisms in our gut, the microbiome. The easiest way to shift the balance of your microbiome from life-serving to unhealthy, is by feeding yourself a lot of sugar. Unfortunately the long term effect of sugar cannot be felt in the short term, we tend to be blind to it. This is used by the food-industry by adding astounding quantities of sugar to everything.

In addition to the sugar, the methods of the food-industry involve using chemicals in order to optimise the yield in agriculture, as well as in livestock farming. This results in humans consuming these chemicals, which act as toxins or hormones - both disrupting the functioning of the body.

In the attempt to get more yield from each square meter, the soil gets exploited to a level that it lacks important nutrients. This results in poor nutrient values of the food we eat.

Next to food, we can disrupt our microbiome with other sources of toxic contamination via the air. During rush hour in a city, your body is not only exposed to the exhaust of countless cars, but surprisingly enough, to the same level of parts per million to chemicals from personal hygiene products.

This industry is another one that has fallen prey to the mechanics of the way we have organised our economic system. It has diverted from helping people to maintain their personal hygiene, to an industry that has created very similar models of dependency and addiction as the pharma- and food-industries.

In summary, these industries are driven by money to such an extent it makes them blind or ignorant to the negative effect on the human body.

In fact, one could argue that the the pharma-industry has no incentive to ensure that we get healthy!

I am not saying that they are doing this on purpose. I know that many representatives of these industries genuinely believe that they are doing a valuable service to humanity. In the trappings of the current system they are doing the best they can. But as long as this continues, they cannot escape the fact that it forces them to optimise the system, instead of the good of the people they supposedly serve.

What it must look like

There is interesting research happening about the workings behind placebo and nocebo effects. As it turns out, the self-healing properties of the body are switched off whenever we go into stress response - commonly known as fight or flight mode - and switched on during relaxation response. This mechanism seems to be responsible for cases of spontaneous remission. There have been documented for almost any type of illness including cancer, HIV and diabetes.

The relaxation response can actively be triggered by a range of things. Dr. Lissa Rankin, a specialist in the field, lists six especially important steps: Believe you can heal yourself. Find the right support. Listen to your body and intuition. Diagnose the root causes of your illness. Write the prescription for yourself. Surrender attachment to outcomes.

This list does not include any standard protocols from modern medicine, but scientific evidence shows that it works. But of course there is more , our health is a function of what we put into our bodies. The food and the thoughts we "consume" create an environment which controls the expression of genes. Some genes we like to be expressed, others we don't.

Most of the genes that are being expressed are not even our own. More than half of the cells in our body are not human and these cells carry their own DNA. In fact, some of them carry a lot more than human DNA.

Most of the foreign cells live in the human gut, doing the work that provides energy to our bodies by breaking down the food we eat. Then it's passed on to the mitochondria in our cells in order to turn the food into energy. This so called microbiome, the micro-organisms living in your gut, can be disrupted very easily by certain types of food. Disruption in this case means that you foster the growth of microorganisms which are bad for you, and in essence will kill you!

Today we know which types of food we should eat in order to feed the right micro-organisms and to kill the ones we don't want - or at least most often. There is a lot of literature available on this for anyone wanting to learn more, so I won't repeat all the science and findings here. But we know that with the right diet can ensure great health and longevity.

Science has verified by now that a plant based diet has incredible health benefits over any other type of diet. The myth that we require animal protein has been debunked time and time again. In addition to the health benefits it is also a lot healthier for the planet when we feed ourselves with plants.

One person can be fed for a whole year with approximately 300 square meters of land. If all 10 billion (2050) humans on the planet eat a plant based diet, this would require around 3 million square kilometres, about 8 times the size of Germany.

While this sounds like a lot, growing food on the vertical axis will reduce this to 20% which leaves us at 600.000 square kilometres. For a city like Berlin, this would mean an area the same size as the city would feed the entire population! This still sounds like a lot, but If you envision this as a ring, it would form a ring around the city only 200 meters wide. That is a little more than seven football fields.

This is why, in addition to dramatically reducing the land used for agriculture, we must also move the fields of indoor and vertical farming forward. It is no problem to grow healthy food in a box on the North Pole or anywhere else in the world. At the same time we are learning how to grow food in the lab that otherwise would require enormous resources. This way we can create types of food that supplement the diet on a very personal and individual level. These are based on your biomarkers, such as your proteomics fingerprint, which describes how your body folds protein at any given point in time. That information can give us important insight into our current health situation. Nutrition that is fitted exactly to what your body needs at any given point in time, long before you develop any symptoms, will change the way we think about health forever.

Of course it is crucial to have a system you can fully trust with this data. After all, this is about your personal health and could be exploited in numerous ways. While there is no silver bullet to solve this, looking at the way Estonia handles this, points - again - in the direction of distributed ledgers. Instead of protecting access to the data, you can make it impossible to do so secretly. Simply create penalties for those who access other peoples data without permission.

Also, treating our personal hygiene in a similar way ensures that the organisms you are hosting take great care of you. After all, their life depends on yours.

Our psychological health also plays a crucial role in our physical wellbeing. We will discuss this in the later chapters: "Relating to People" and "Minding the Spirit".

To sum up, all of this sounds too easy to be true - but it is. If you want humanity to survive this century and beyond, doing everything in your power to move towards a plant based diet is key for healthy humans and the planet.

A lot of this knowledge is still alive and active in many ancient traditions around the globe. Identifying, preserving and spreading this knowledge from the islands where it currently lives, can be fostered and supported by technology. Creating a marketplace of nutritional, agricultural and medicinal best practices, that allows marginalised ideas and people to take part in the economy and profit from their valuable knowledge, is very doable.

Once formalised in a digital system, preferably one without a central stakeholder, but a system owned by the general public, the knowledge might also be used and accessed by machines. Humans could be informed of best practices that would solve any constraints in real time.

Minding the Spirit

Why we do it

Etymologically the word spirit is closely related to "Breath". In most religious systems it is used to describe a higher being, sometimes synonymous with soul , something that does not end with the death of the physical body.

   In this context I use the term to describe something that is neither the body, nor the mind - but a third entity in its own right. Something even harder to grasp than mind.

Just like the brains, the clusters of neuronal cells we have in our bodies - not only in our head, but also around our heart and gut - can be seen as the interface between the body and the mind, I see consciousness as the interface between the mind and the spirit.

I use the term consciousness to describe the level of mindset evolution a person has reached, building on Robert Kegans' framework which I mentioned earlier. To recap:

Basically, there seems to be three layers of development of the human mind. It begins with the "socialised mind" which is mostly influenced and defined by its immediate social surroundings. This enables a person to be part of a community and to be trusted by members of that community by acting in predictable ways.

It is followed by the "self-authoring mind" which begins to look at the memes that are used in the community in order for it to work. And also to make conscious decisions about which ones a person wants to follow, which to ignore and note those which could be useful.

And then, finally, the "self-transforming mind" is able to step out even further. It begins to look at its own mind and behaviour and make decisions about it, essentially creating its own transformation.

The spirit is not the same as consciousness, just like the brain is not the same as the mind. Just like the brain is the substrate that the mind runs on, consciousness arises from the complexity of that mind and serves as the substrate for the spirit.

This does not mean that the spirit evolves with the consciousness. Seen from a different perspective, it could be defined as the essence of a person. This essence is always there, but it gets more refined and more pronounced with the individual evolution of consciousness.

The spirit in this definition expresses in the intuitive and authentic everyday action and non-action of a human. I will use the same perspective during the exploration of the spirit.

How we do it (and what goes wrong)

Currently we go to one of two extremes when minding the spirit. We entirely ignore it or we blow it out of proportion. I will explore both, starting with the former.

The Newtonian age has offered us a great deal of understanding about physical reality. This understanding has led to enormous achievements. It has also led us to focus entirely on things that can be calculated, measured, touched or seen.

Even though we don't exactly know what it is or how to describe it, we feel that something is missing. We know that full immersion in a lifestyle built on the Newtonian mindset is not fulfilling, leaving something behind, ignoring an important component we cannot name.

The truth is, it has a name - the name is "The Spirit". Unfortunately the connection to this name is not satisfactory for many people today because they have been massively blown out of proportion.

Before we started making systematic and scientific sense of the world, things that scared us, things that could not be explained could - very conveniently - be ascribed to a higher force, a being, a ruler of life and death, a creator of earth. Or sometimes just a part of life as we know it. In one form or another, this god, or group of gods, had been associated with spirit.

This paradigm worked for a long time in human history for explaining things and creating order by imposing rules and guidelines of how to live a good life. It has also created a lot of evil for the people who took a different perspective. An important element of any successful religious meme is to be missionary, which often turns into a forceful way of imposing the ideas onto a populace.

Today we live in a world that is still far from being understood or explained by modern science. Large portions of the understandable universe are still open for assumption and exploration. This creates new room for spiritual belief systems to explain the parts that science cannot, or has not yet explained - and maybe never will.

Neither modern science nor any spiritual path has definite answers to unanswered questions. You are either forced to wait with your knowing until science has figured something out, or choose to believe something that might turn out to be non-sense, or not.

Humans seem to have a hard time holding the space around the insecurity of not knowing, without becoming fundamental in their beliefs. We take action, we say and think things that are incoherent with our own understanding of the world. Why? Because we carry conflicting views with no method of resolving them.

What it must look like

Despite our subjective experience of being in control of our life and decisions, 95% of our time we run on autopilot.

As Daniel Kahneman reported in his book *"Thinking fast and slow"*, our brains prefer using our limbic system over using the pre-frontal cortex because it is much more energy efficient and also a lot slower than the limbic system. In fact the pre-frontal cortex is one of the most energy hungry parts of the body and limiting its activity is crucial for the brain.

The pre-frontal cortex is the home for our conscious thinking, like planning, interpreting, and structuring, whereas our limbic system is the home for our subconscious mind. Our subconscious mind operates on heuristics. To be able to generate quick reactions to the many inputs we are exposed to throughout the day.

These heuristics got there mostly by repetition, especially during the first couple of years of our lives, when the people and society around us explained their view of how things work in the world.

In doing so, they effectively programmed you to be a functioning member of society - putting you in the state of Kegans' socialised mind.

Operating on this level for an entire life is absolutely possible but many people find during the course of their life that there is some misalignment between how they consciously perceive the world around them and the memes that were implanted in their subconscious. Once an individual realises this and starts deciding which of the memes she wants to use and which she would rather not use, advances her to the level of the self-authoring mind.

Her next step is not only watching societies rules from outside of the system, but outside of her own system - the contents of her subconscious. Doing this with her conscious mind and then editing and designing its contents, elevates her to the self-transforming mind.

This is not only possible but it also can be trained. Many ancient traditions have developed tools to precisely achieve this state. While these tools might differ in their approach, their objective is identical. They are like paths leading up the same mountain on different routes. No matter which one you take, it will bring you to the summit.

Based on these traditions it is possible to design programs that are in tune with the cultural environment an individual finds herself in order to help develop a self transforming mindset.

As I mentioned earlier, this is not new. When it became evident in the nineteenth century in Germany that there would be a massive shift in the way society was organised, the leading idea was that this shift would be moving power away from a few individuals in the direction of the people. The circles around the leading thinkers of the time, like Goethe and Schiller, developed the concept of providing retreat camps for young citizens. The ideas was to help them find their center and ground themselves as a basis for creating such a people-driven society.

The revolutions around 1848 stopped these ideas in their tracks but they were picked up half a century later in the nordics as outlined in *"The nordic secret"*. Similar to what the nordics did by providing unconditional free retreat centres to young people, we can do the same on a global level. Using technology to facilitate access to the tools and methods required; then create and facilitate mentoring support.

These procedures need to be highly transparent and and freely accessible to everyone on the planet. Also, the level of personal development can be documented in a way, that allows everyone to have a more in-depth understanding of the underlying reasons for reactions and actions of individuals. Much in the way large corporations today give their employees systems like HBDI, or similar, to achieve the same end.

Imagining a Good World

Now that I have sketched some thoughts around six dimensions of the human condition and society, it is time to consolidate this into one coherent and understandable picture. A picture that is desirable, feasible and doable.

As I set out on this endeavour, I am filled with anxiety and doubt. There are so many people out there on the planet, who am I to come up with a solution for all of humanity? The answer or course is, that I am not. I am putting the first strokes onto a canvas that can be painted by many, maybe one day by everyone working in harmony.

So, what follows is only a first layer of paint on a blank canvas. Take it as an invitation to build upon it, change it, use it and move the effort of creating a society that can both restore the planetary balance and thrive in this balance. With that in mind, let me begin.

Whether some people decide to live secluded in a small group or alone, or whether they prefer larger groups, what everyone will have in common is that their homes will be essentially entirely off-grid and built with materials that can be found mainly at the location of construction. Earth, sand and clay provide a solid material for building. These buildings will not stretch up into the sky, but will be small buildings, inhabited by individuals or small groups, like families. Allowing a generous 50 square meters per person for a population of ten billion people, the total area required for the entire population of the world would be less than one and a half times the area of Germany. When spread out over a large area, these dwellings can exist entirely off-grid using current (2019) technological standards.

If we add to this the space we need for feeding ten billion people using five levels of off-grid vertical farming, including hydro- and aeroponics, we would only need another area a little more than one and a half times the area of Germany.

Let's add some space in-between for transport and recreation. The number of meters of paved road per German citizen in Germany is 8 meters. Using this measurement we can allot a generous 100 square meters per person for transport. This would require another three Germanies. Adding

30% of the space used for housing and transport for recreation, we add another generous one and a half Germanies.

In summary the total area required to house and feed an entire population of ten billion people would fit into an area seven and a half times the area of Germany, which is a little over half the area of the entire EU.

So, in the future we can live in off-grid houses much like the concept of earth ships developed by Michael Reynolds. They will be built with sustainable material and essentially be carbon negative. We can design communities of varying sizes, but they will have a few things in common. A good model is the one developed by Jaques Fresco in the Venus Project. The more commonly shared and social functions of the community are closer to the center and the other elements like housing, recreation, energy creation, food creation are arranged in circles around this center. Each circle is wide enough to provide the required space for its function. Just as a reminder, the food ring for a city like Berlin would be five stories high and two hundred meters wide, providing food for the entire city. While it is possible to have such large cities, most of them will be much smaller. Many cells of circular villages could make up a larger city, but each one is self-sustaining and self-organising.

Each of the cells provide all the food for the inhabitants in vertical farms using aero- and aquaponics and are capable of producing almost any climate while being entirely off-grid. The land surrounding the cell can also be re-wilded, providing space for flora and fauna to reclaim the space.

We will support this by planting redwood, hemp and other plants with high coefficients of carbon sequestration, in order to pull as much CO_2 out of the atmosphere as quickly as possible. Turning the entire agricultural space of the planet back into wild nature will sequester roughly 17 billion tons of CO_2 per year. In combination with sharply reduced emissions, this will help bring the carbon cycle back into a balance that is maintainable for nature.

The reduction of emissions will be mainly achieved through the elimination of transportation since food is produced close to where it is consumed. Another factor is the transition into a fully vegan diet. To support this we will invent entirely new categories of delicious plant based food which builds on our inherent need for social interaction and shared meals. Being creative in food preparation will be one of the ways of receiving positive social recognition. But this is not limited to food related creativity. Being able to express complex feelings and situations in innovative and creative ways, through art forms like music, dance, theatre, sculpture, or future art forms will be highly regarded. These all play a vital role in creating a consensus within a community and allowing people to heal chronic disease through new forms of expression. Especially those helping them to get in touch with their authentic self and articulating their needs.

While the elements described so far sound like going back to a tribal culture, there will be another layer. A layer of technology, connecting all of these cells, weaving them together into a mesh of participants in a global conversation. Discussing the state of the planet, the index of available resources and their best use. This works through a distributed network for computation and data storage making it highly resilient against physical disruption and impossible to control by any single entity. The computations on this network will be translating the sensory input received from the city- and village-cells and mapping them onto the available resources. In other words, they act as a global ERP system that knows supply and demand and has heuristics to make decisions based on the inputs it receives.

Managing these heuristics will be the centrepiece of what we call "work" in this world. We will need to engage in a constant dialogue and discourse about the needs of our local community, as well as the best potential use of the scarce resources. Making these seemingly difficult decisions will be a lot easier than today, since almost every need of the people will be supplied by local produce.

This process will be enabled by adopting the original form of democracy from ancient Athens, into a modern form of multi-body, sortition-based democracy, as put forward by Terrill Bouricius. In this form, several bodies of government create a solid system of checks and balances, as well opening the possibility for participation by entities not currently taking part in the political process. The people engaging in these bodies of government are not voted for, but instead are chosen randomly and assigned for a specific period of time. This kind of civil service will become a central part of our work in the future.

However this type of organisation only works on a local level. We will need to create a similar structure on a global level, feeding back local decisions at high speed. This will most likely be achieved using brain machine interfaces, in order to keep up with the speed of the machines and the vastness of the global conversation. This part of the global political process will be a blend between man and machine. Using the combined conversation of the decision makers from local communities will form and shape the heuristics used by the machines in real time, while also feeding the data used as a basis for decision making.

These global decisions will revolve mostly around shared and global means of transportation as well as rare earths where still required to create high tech products to facilitate this global network from a computational and connectivity as well as from an energy production perspective.

The devices required to serve these needs will not be manufactured in large central factories, but in local multi-purpose, micro-factories. These will be able to produce almost anything. Utilising a mix of robotic arms to format and move things, 3D printing techniques for any material imaginable, plus using synthetic biology for growing materials and products in the lab.

Adding to the two important forms of work of cultural contribution and political participation, a third core activity will be scientific work. For example, creating new and better ways to harvest and store energy, balance the carbon cycle and other planetary ecosystems, as well as improving computing both on a hard- and software level. Similar to the other two domains, much of this work will be performed with the help and collaboration of machines working at incredible speeds, using mind-machine interfaces.

A fourth dimension of the work of the future revolves around caring for other people. This function replaces and embraces the current forms of the kindergarten, school, hospitals, old age homes and similar institutions. I want to call this function The Supporter. The Supporter provides guidance and help to humans in their various stages of life. Often The Supporter does not know how to help him or herself but will make sure that help will be provided either by machine or human, depending on the situation.

The early 21st century school-teacher will be entirely obsolete, since machines will be better at providing learning material and helping people turning information into knowledge on an individual level, than any human possibly could. So the role of The Supporter in this case will be able to help guide the young human to find his or her fields of interest and open the possibilities of the world to this person, rather than copying information from her own brain into that of a child.

Since all of the data storage and computation happens on globally distributed networks and all human activity is recorded on these networks, the people who are highly active and successful in these domains will be easy to identify. A culture of celebration of these global heroes will replace the current ways of recognising social status, based on material consumption. That will have totally vanished!

As described there will still be work corresponding with Maslows' hierarchy of needs. Everyone will spend equal amounts of time during their waking hours nurturing nature, their individual, local and cultural expression.

Maybe your mind rejects these ideas. Usually this happens with thoughts like: "Ya, right - Dude! - Shat do you smoke?" I get it. The next chapter is for you. It is the entire reason for creating this thought experiment and putting in many hours to present this weirdness in the hope that we can make it through this century with ten billion thriving human beings.

Building It

Nothing is harder than changing a system from within. Earlier I talked about the two main factors that are in gridlock in our society, causing all the harm we are doing to the planet and the people:

- the design of our currency system and
- the current configuration of our democracies

I think and work from the assumption that everyone involved simply wants to do the best they can, so let us not indulge in finger pointing and guilt-games. Let us focus on how we can fix things.

Based on the systemic gridlock I don't believe that it is in any way possible to change the system at all. We will not change our backing system and we will not significantly change our democratic make-up. The forces of inertia are simply too strong.

So, instead of changing the system we will simply ignore it and build a new one. We will start by encrouraging a small number of people to log out of the current system. Then we make some components of their lives so much better, that other people will start to follow.

We will do this in a way that it is easy for people with an entrepreneurial spirit to adopt working models for their own environment.

Repeating this process we will eventually reach a critical mass of people and functions that have simply been removed from the current system.

Creating small, local cooperatives based on distributed ledger technology will cover one aspect of life and allow the people taking part in the cooperative to remove themselves from the current way of doing things.

I will now sketch some concrete examples of what this can look like. I believe these must be implemented, although I am not providing entire implementation blueprints at this time.

Farming

Based on current technology, including ancient knowledge about building houses, it should be possible to build off-grid vertical farms. In terms of building technology these can easily be modelled after the earthship concept of building sustainable off-grid housing.

There is a sweet-spot between the size of the farm, price of construction the and maintenance that allows for a group of people to pool the necessary resources for buying the land, then building and running the facility.

The rights to participate at minimal or no cost is documented in the distributed ledger which was used to collect the funds in the first place.

Based on this model, farms all across the globe can use the same software infrastructure to build their cooperative, as well as using the existing blueprints.

The original authors and investors can hold tokens that appreciate over time as the total value of the entire ecosystem grows. This is described in "Cyptoassets" By Chris Burniske and Jack Tatar, who liken the token-system around a specific commodity to an economy which allows to calculate the GDP of this economy. The growth of GDP reflects the increase in value of the system and hence the increase in the financial value of the initial tokens.

It's worth noting, for some of the projects to kickstart a new economic model, it will be necessary to build on the incentive systems of the old model, even if these become obsolete later.

Housing

Much like farming, housing depends on ownership of land and the off-grid capability of the houses themselves.

In a similar fashion to the farming model, a group of people can pool money on a distributed ledger to buy land and build off-grid houses in the model of earth ships. Or use the Zappos financed startup Geoship, or similar models.

The ledger originally used to pool money will also serve as a registry of ownership. This can also be used to swap housing with other people across the globe, through a network of cooperatives building on the same housing-ledger.

Transport

Transport can be handled in a similar way. However, there are some complexities that need to be accounted for. Transport of people can be tied in with the Housing-Cooperatives, while being built in parallel.

In these systems, the vehicle of choice, like scooter, bike, e-bike, e-car or drone, is acquired by the cooperative and paid off in a model like the existing sharing services. While participants in the cooperative receive special prices, the general public is also invited to use the vehicles, at a higher price-point. Once the purchase price of the vehicle is paid, the earnings of each vehicle are split across the cooperative.

These credits could also be used in other transport cooperatives for short, mid or long range transport.

Regarding transport of goods, a lot of work and research has been conducted and we have a good understanding of the most efficient ways of transporting goods around the globe. In essence, it is a hub-and-spoke network similar to the way we have constructed the internet to be resilient.

One existing solution on the horizon is Matternet. This is using the concept of Matternet and converting it into a token-economy in combination with hardware for transport, maintenance, parking/landing and charging. Then, with the right incentives within the design of the token-economy, it has the power to not only spread quickly, but also ensure that the future of the transport system is a commons. This is better than being a product of a company, or owned by one or more governments.

Health

If you follow the money, it becomes evident that the powerhouse of the health-industry is the insurance companies. If an insurance company does not accept a certain drug, it is almost impossible to put it into the market.

Given the nature of distributed ledger technology it is very straightforward to implement an insurance model based on this technology. But it must not stop there. Intelligent token engineering has the potential to distribute the entire risk-assessment and administrative processes, into the crowd as part of the incentives in the system.

Paired with the collection of personal health and fitness data on a system that is neither owned nor cannot be controlled by a single entity, this can be turned into a powerful driver towards true preventative health care. Essentially this turns the dystopian idea of an all-controlling state, or company in this sector, into a utopian version of the data that is owned and controlled by the entire community of participants.

Clothing

We cannot ignore something as elementary both from a physical as well as from a psychological perspective as clothing.

If you look at the balance sheets of the large corporates in the fashion industry, an interesting trend emerges. All of them use roughly 50% of their revenues for marketing. Redirecting this 50% into making the value chain of the product truly sustainable.

The example of Primark shows that it is possible to rethink the idea of marketing. In essence, Primark outsources the marketing activity to the customers, rewarding them with lower prices than the competition, effectively paying them for the marketing.

By creating a system of local, and even exclusive brands, around individuals in horizontal as well as vertical communities, it becomes possible to create the desirability we are used to when it comes to clothing. This dramatically reduces the marketing cost and also slows the fashion cycle to a healthy and more sustainable speed. Adding structured secondary markets to these brands would increase the sustainability even more. For example, using automated processes and alternative materials like hemp, which has the added benefit of sequestering a lot of carbon from the atmosphere.

It is therefore possible to create a platform that manages small and local brands and their entire value chain. While this can be imagined as a centralised platform, it can be decentralised through clever incentive mechanisms for sourcing materials, providing designs, creating the product, etc. This way it becomes much more resilient, as well as adaptable to local and cultural requirements and can scale faster.

Knowledge

All of this will not be possible without a strong and solid foundation and agreement on the globally accepted knowledge in a way that can be accessed and shared easily.

Again, it is very helpful to imagine this in a distributed way. Knowledge is power. Being a gatekeeper to important knowledge in the world provides a lot of power. This power must be a commons.

There are several experiments in this space at the moment. For example civil, a distributed platform for journalism. They use an incentive model that rewards truthful evaluation of content. Whether their specific model works and can be operated sustainably, remains to be seen.

The idea of incentivising the search for truth in information and knowledge in a way that cannot be gamed by single entities is unique. This is a more democratic process that is already being explored.

All that is needed is the energy of an entrepreneur, or a group of people, who can build an incentive mechanism of this kind into a token. This would create a truly global platform for managing the knowledge of humanity.

What's missing...

I could go on forever about other fields of life and the world, like culture, network hardware, computational hardware, software, etc. But neither do I want to bore you with these, nor is it necessary, because the general picture is already emerging very clearly from the previous examples.

You can take any existing product or value chain, rethink the incentive mechanisms and turn it into a distributed software product. This is owned by the participants and the initial invest gets returned through owning part of the transaction power within this network.

Next steps

We can achieve all of this by creating a link between the entrepreneurs who are capable of seeing and building this future, and the investors who are capable of seeing it and financing it.

My personal next step will be to invite you, or people you know, who fall into one of these categories to participate in a series of conversations. These will be centred around building a new operating system for the way we live on our planet - Mother Earth.

Thank you for making it all the way here! I am very much looking forward to hearing from you.

Thomas

t@thomas.cr

+491785830082

I am deeply grateful

to all the people who have made this book possible by giving me time, trust, love, ideas, feedback, strength and criticism. You know who you are.

Especially I want to thank my friends Tom, Frank, Lars, Sebastian, and Manuel for their patience and support. And Les for making it a lot more readable!

There are a few people without whom this book would not have been possible at all. These are Barbara; my beloved mother, Vanessa ; my cherished partner and Noah; my inspiring son. Thank you!

Further Reading

- Frederic Vester - The Art of Interconnected Thinking
- Yuval Noah Harari - Sapiens
- Yuval Noah Harari - Homo Deus
- Yuval Noah Harari - Lessons for the 21st Century
- Don Miguel Ruiz - The Four Agreements: A Practical Guide to Personal Freedom
- Daniel Kahneman - Thinking Fast And Slow
- Carol Dweck - Mindset
- Ray Dalio - Principles: Life and Work

- Sam Harris - Waking Up: A Guide to Spirituality Without Religion
- Mihaly Csikszentmihalyi - Flow: The Psychology of Optimal Experience
- Thomas Harris - I'm Ok, You'Re Ok
- Harry G. Frankfurt - On Bullshit
- Michael Puett - The Path: What Chinese Philosophers Can Teach Us About the Good Life
- Nick Bostrom - Superintelligence
- Max Tegmark: Life 3.0
- Geoffrey G. Parker, Marshall W. van Alstyne, Sangeet Paul Choudary - Platform Revolution: How Networked Markets Are Transforming the Economy and How to Make Them Work for You

- Chris Burniske, Jack Tatar - Cryptoassets: The Innovative Investor's Guide to Bitcoin and Beyond
- Charles Eisenstein - Sacred Economics
- Bernard Lietaer - The Future Of Money
- Jeremy Rifkin - Access
- Peter Thiel - Zero To One
- Frederic Laloux - Reinventing Organisations
- Salim Ismail - Exponential Oranizations
- Steven Johnson - Emergence: The Connected Lives of Ants, Brains, Cities, and Software
- Andrew Grove - High Output Management
- Susan Pinker - The Village Effect
- David van Reybrouck - Against Elections

- Lloyd Sieden - Buckminster Fuller's Universe: An Appreciation
- Shimon Peres - No Room for Small Dreams: Courage, Imagination, and the Making of Modern Israel
- National Research Council and Committee on Twenty-First Century Systems Agriculture - Toward Sustainable Agriculture: Agricultural Systems in the 21st Century
- Steven Gundry - The Longevity Paradox
- Liliann Fischer, Joe Hasell, J. Christopher Proctor, David Uwakwe - Rethinking Economics
- Donella Meadows - Limits To Growth: The 30 Year Update
- Scott Carney - What doesn't kill us

- Richard Brodie - Virus of the mind
- Ballot Eric, Meller Russell, Montreuil Benoit - The Physical Internet
- Lawrence Lessig - America, Compromised
- Paul Gibbons - The science of successful organisational change

www.ingramcontent.com/pod-product-compliance
Lightning Source LLC
Chambersburg PA
CBHW030647220526

45463CB00005B/1667